MW01318774

# The Great Ocean

by Jennifer Waters

Content and Reading Adviser: Mary Beth Fletcher, Ed.D.
Educational Consultant/Reading Specialist
The Carroll School, Lincoln, Massachusetts

Minneapolis, Minnesota

Compass Point Books
3722 West 50th Street, #115
Minneapolis, MN 55410

Visit Compass Point Books on the Internet at *www.compasspointbooks.com*
or e-mail your request to *custserv@compasspointbooks.com*

Photographs ©: DigitalVision, cover; TRIP, 4, 6, 14, 16; PhotoDisc, 5, 8, 9, 11, 13, 15, 17, 19; Dave G. Houser/Corbis, 7; Eyewire/Getty Images, 10; David Muench/Corbis, 12; TRIP/G. Hancock, 18.

Project Manager: Rebecca Weber McEwen
Editor: Heidi Schoof
Photo Researcher: Image Select International Limited
Photo Selectors: Rebecca Weber McEwen and Heidi Schoof
Designer: Erin Scott, SARIN creative
Illustrator: Anna-Maria Crum

**Library of Congress Cataloging-in-Publication Data**

Waters, Jennifer.
The great ocean / by Jennifer Waters.
v. cm. — (Spyglass books)
Includes index.
Contents: Earth's great ocean—Pacific Ocean—Atlantic Ocean—
Indian Ocean—Arctic Ocean—Ocean life—Did you know—Glossary.

ISBN 0-7565-0379-5 (hardcover)
1. Ocean—Juvenile literature. [1. Ocean.] I. Title. II. Series.
GC21.5 .W38 2002
551.46—dc21

2002002558

# Contents

# Earth's Great Ocean

When you think of planet Earth, you may think about lots of land. Really, most of Earth's surface is covered in water.

## Did You Know?

Most of Earth's water
is found in the oceans.

Oceans are big bodies of ***salt water***. All oceans are connected to each other. Together, they make up "Earth's great ocean."

## Did You Know?

A sea is a small area of ocean that is close to land.

Earth's great ocean has four main bodies of water: the Pacific Ocean, the Atlantic Ocean, the Indian Ocean, and the Arctic Ocean.

1. Pacific Ocean
2. Atlantic Ocean
3. Indian Ocean
4. Arctic Ocean

# Pacific Ocean

The Pacific Ocean is the biggest and oldest ocean. The deepest spot in the world is in the Pacific Ocean. It is almost seven miles (11 kilometers) below Earth's surface!

An ***island*** in the Pacific Ocean

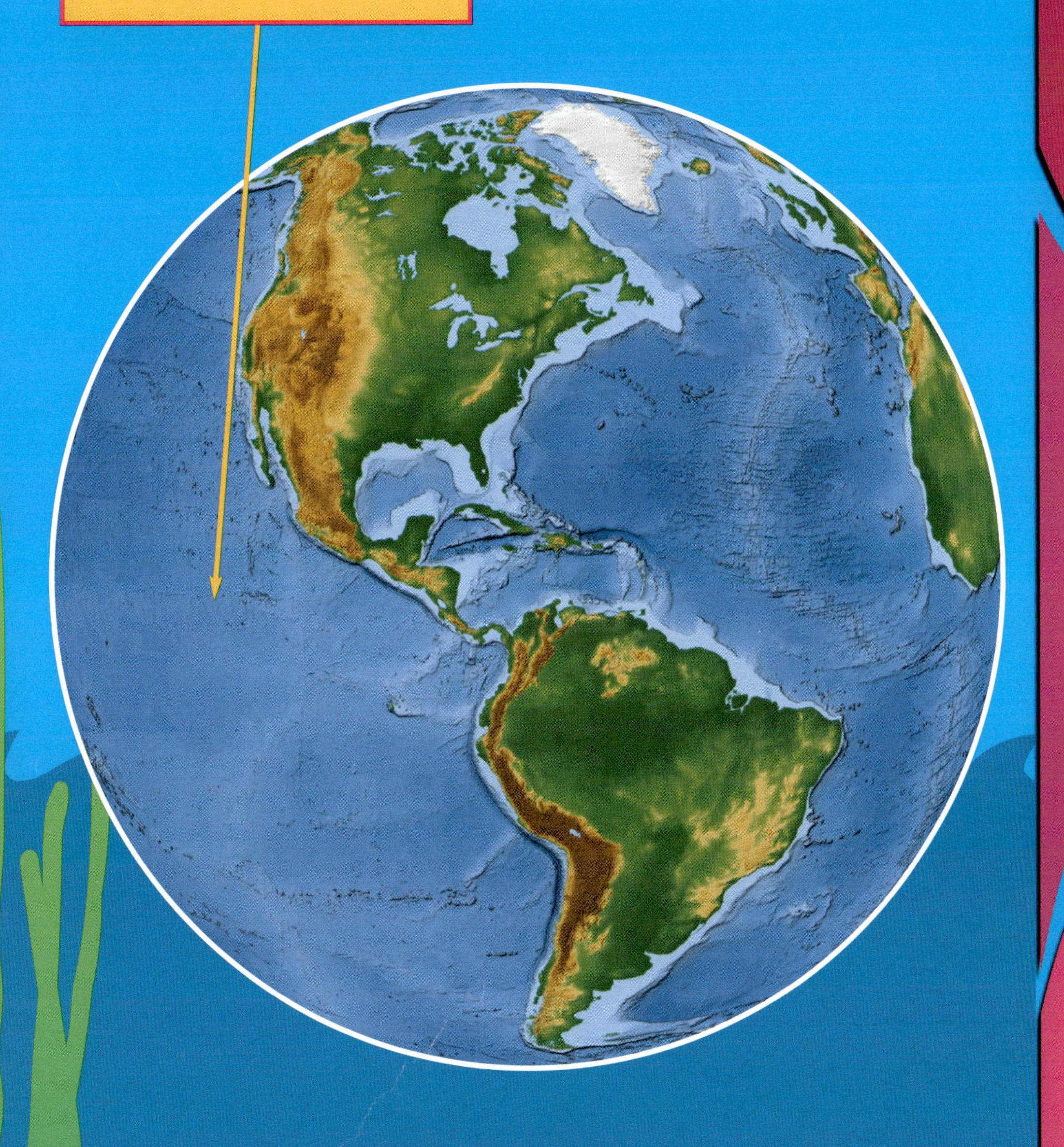
Pacific Ocean

# Atlantic Ocean

The Atlantic Ocean is smaller than the Pacific Ocean. It is the youngest ocean. The name “Atlantic Ocean” comes from Greek ***mythology***. It means “Sea of Atlas.”

A ***lighthouse***

Atlantic Ocean

# Indian Ocean

The Indian Ocean is smaller than the Atlantic Ocean. This ocean has thousands of islands. The Indian Ocean was named by explorers who crossed it while trying to reach ***India***.

Islands in the Indian Ocean

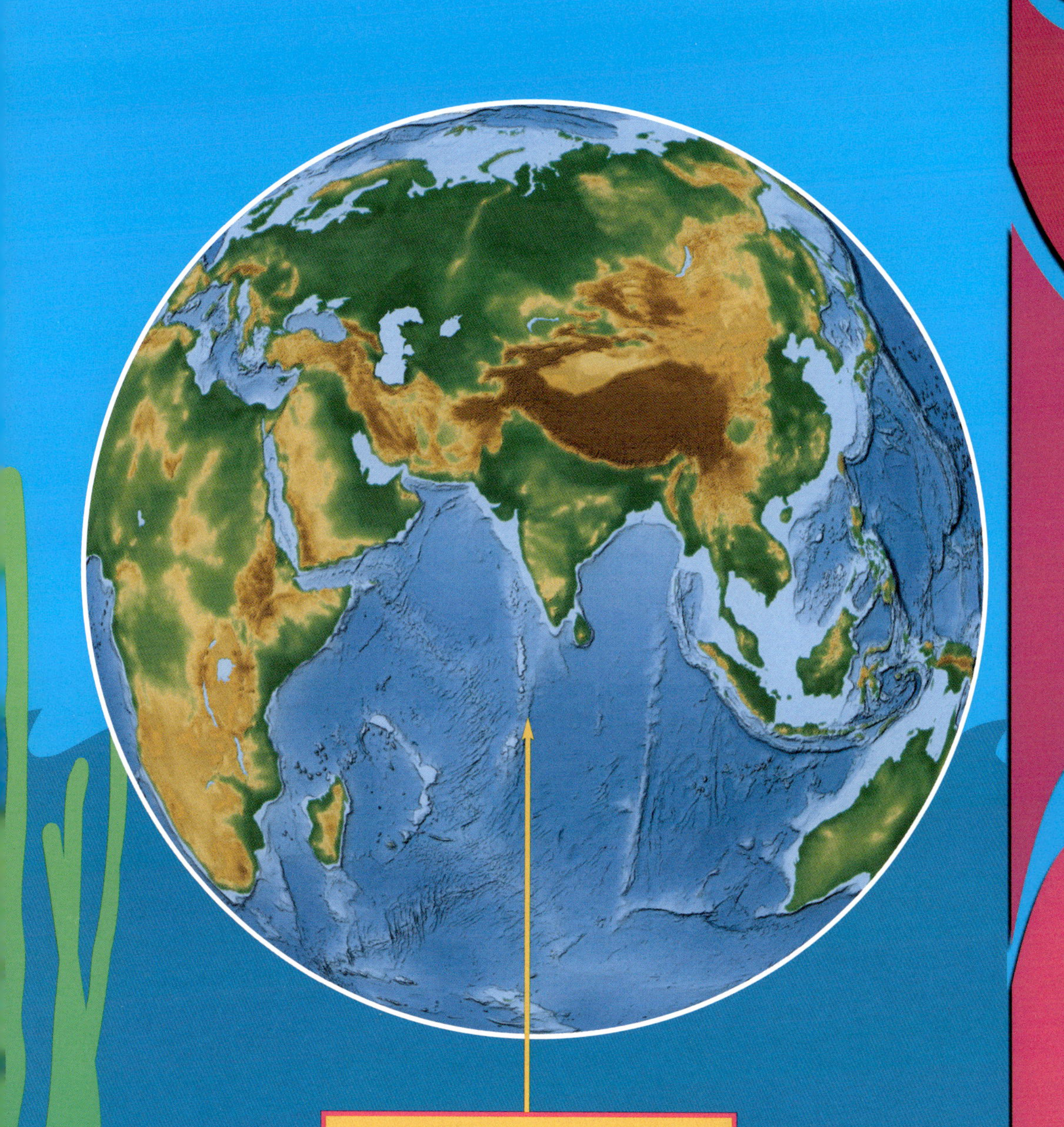

Indian Ocean

# Arctic Ocean

The Arctic Ocean is the smallest and coldest ocean. Ice covers the top of this ocean. Cold winds that blow from the Arctic Ocean cause rain and snow in faraway places.

Arctic icebergs

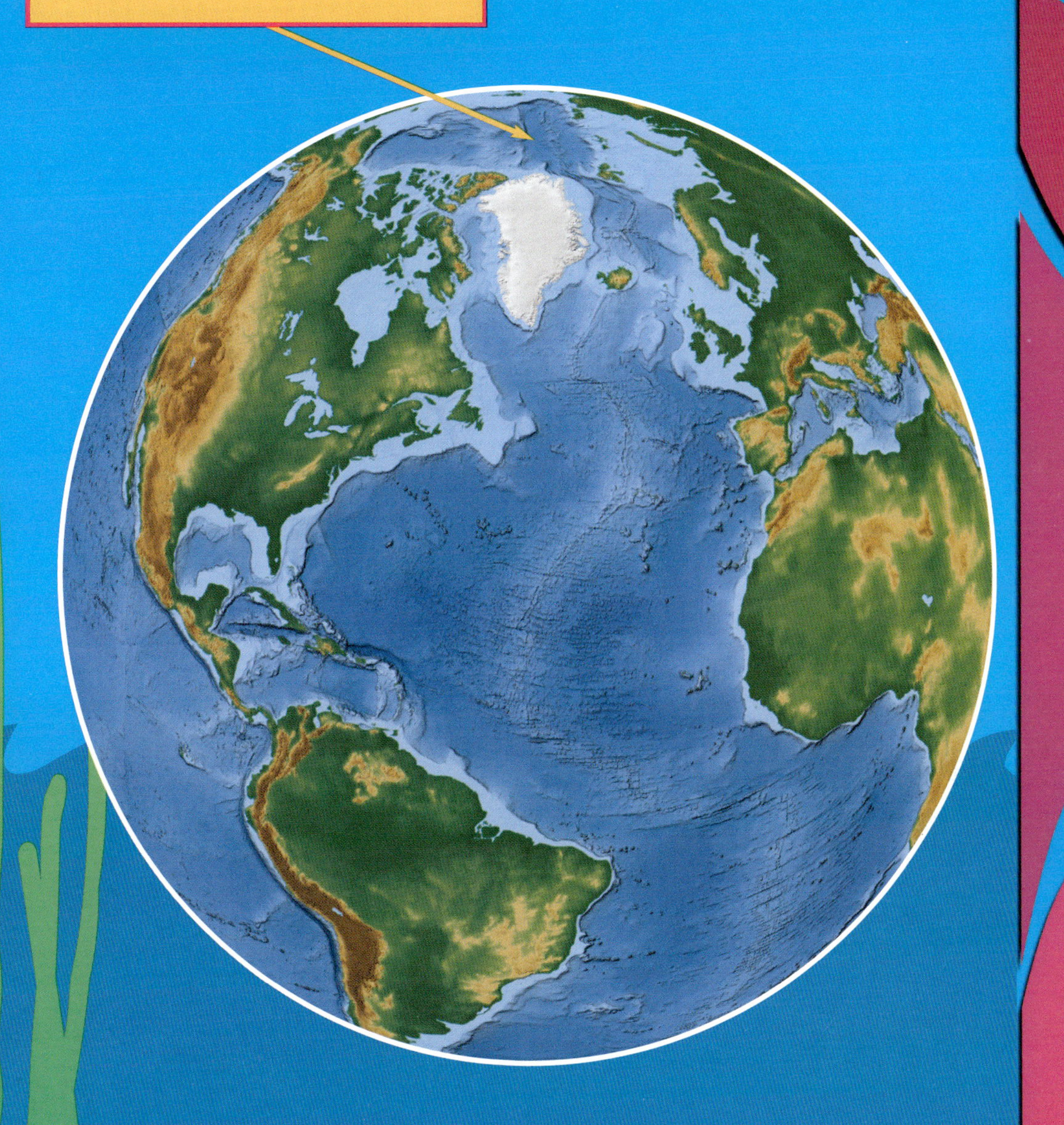
Arctic Ocean

# Ocean Life

The ocean is home to many different plants and animals. ***Pollution*** hurts and kills many creatures that live in the ocean. That is why it's so important to keep our oceans clean.

Polluted water

# Everyone Can Help

Even if you don't live anywhere near water, you can help keep the ocean clean.

1. Remember that wind or rainwater can carry garbage to streams and rivers. Don't litter!

2. Don't just throw away plastic holders for six-packs of soda. Cut the plastic loops into pieces so an ocean animal can't get caught in them.

3. Don't pour paint or oil into ***storm drains***. They could travel into rivers or oceans and make plants or animals very sick.

4. Tell your friends how to help keep the oceans safe and clean.

# Glossary

*India*—a country on the continent called Asia

*island*—an area of land that is surrounded by water

*lighthouse*—a building that flashes a light at night so that ships don't come too near the land and crash

*mythology*—a group of myths, or stories, people tell to explain things about the world

*pollution*—waste that people put into the water, land, and air

*salt water*—water that contains salt

*storm drain*—a drain in the ground where water goes when it rains or snows

# Learn More

## Books


Burnie, David. *Seashore*. New York: Dorling Kindersley, 1994.

Cohen, Caron Lee. *How Many Fish?* Illustrated by S. D. Schindler. New York: HarperCollins, 1998.

Fowler, Allan. *It Could Still Be a Lake*. New York: Children's Press, 1996.


## Web Sites

enviroliteracy.org/students index.php

aqua.org

# Index



**GR: I**

**Word Count: 215**

## From Jennifer Waters

I live near the Rocky Mountains,
but the ocean is my favorite place.
I like to write songs and books.
I hope you enjoyed this book.